Earth's Hidden Wonders: Unveiling the Planet's Most Mysterious and Astonishing Secrets

Preface

From the towering forests to the deepest oceans, Earth holds a wealth of secrets waiting to be explored. As we go about our daily lives, much of the planet's natural beauty and incredible diversity remains hidden, unknown to most of us.

In writing this book, my goal was to shine a light on the unusual and the obscure—to uncover the environmental mysteries and wonders that often go unnoticed. By delving into the lesser-known aspects of our planet, I hope to inspire a deeper connection with the natural world and encourage curiosity about the ecosystems that surround us.

Earth's Hidden Wonders is not just a collection of facts; it's an invitation to explore and appreciate the marvels of our environment.

This book is a tribute to the awe-inspiring resilience of life on Earth and the incredible adaptability of species in the most extreme conditions.

I invite you to embark on this journey with me and discover a side of the planet that few have experienced. The more we learn about these wonders, the more we can appreciate the importance of preserving our world for future generations.

Introduction

Introduction: Exploring Earth's Hidden Wonders

Earth is a place of mystery, wonder, and sometimes, bewilderment. For all our scientific advancements, much of our planet remains unexplored, misunderstood, or simply too bizarre to fit into what we think we know. This book is dedicated to those wonders—the strange, the obscure, and the awe-inspiring phenomena that make Earth such an incredible place to live.

Our journey will take us through extreme environments, reveal the secrets of little-known species, and explore the rare and extraordinary forces of nature that shape the world around us. The natural world is filled with surprises, and often, the more we look, the more we realize just how much we have left to discover.

I hope that as you turn these pages, you'll find yourself not only intrigued by the strange facts and stories but also inspired by the beauty and complexity of our shared home. Let's begin this adventure and see where the hidden wonders of our planet will lead us.

Chapter 1: Unique Ecosystems

The Atacama Desert: The Driest Place on Earth The Atacama Desert in Chile is so dry that parts of it have never recorded rainfall in human history. Yet, life has adapted to survive in this harsh environment. Scientists have discovered microbial life thriving in soil that seems inhospitable to any form of existence. These microbes, living beneath the surface, rely on the moisture trapped in fog for survival, making this desert one of the most extreme examples of life's resilience.

Deep-Sea Hydrothermal Vents: Life Without Sunlight Beneath the ocean, in the darkest depths, hydrothermal vents spew mineral-rich water heated by the Earth's core. These vents create unique ecosystems where life thrives without the need for sunlight. Bacteria feed on the chemicals released by the vents, forming the foundation of a food web that supports giant tube worms, clams, and other strange creatures. These deep-sea communities challenge everything we know about life's dependence on sunlight.

Antarctica's Dry Valleys: A Polar Desert Like No Other While Antarctica is known for its ice and snow, the McMurdo Dry Valleys are a cold, ice-free desert. These valleys are some of the most extreme environments on Earth, with low humidity and high winds. Despite the harsh conditions, life has found a way here too—microbes live in the soils and in the ice-covered lakes that dot the region.

The Amazon's Underground Rivers: Nature's Hidden Waterways Beneath the lush Amazon rainforest lies an underground river system that mirrors the massive Amazon River above it. This subterranean waterway flows slowly, with water temperatures that reach boiling in some areas. The discovery of these underground rivers has expanded our understanding of how water shapes the planet's ecosystems.

Chapter 2: Rare and Extreme Weather Phenomena

Fire Tornadoes: Nature's Fiery Vortex When wildfires rage, they sometimes create a rare phenomenon known as a fire tornado. These spinning vortices of flame are created when intense heat and turbulent air currents combine, forming a whirlwind of fire that can reach several hundred feet into the sky. Fire tornadoes are both awe-inspiring and terrifying, reminding us of the power of nature in its most destructive form.

The Catatumbo Lightning: Venezuela's Endless Storm In a small area over Lake Maracaibo in Venezuela, lightning storms occur almost every night, sometimes for up to 10 hours at a time. This phenomenon, known as the Catatumbo lightning, is caused by a unique combination of atmospheric conditions, creating the most concentrated lightning storm on Earth. The flashes of lightning can be seen from over 250 miles away and provide a natural light show like no other.

Volcanic Lightning: A Clash of Fire and Electricity When a volcano erupts, it can generate powerful electrical charges within the ash plume, leading to lightning storms in the sky above the eruption. This phenomenon, known as volcanic lightning, occurs when particles in the ash cloud collide and create static electricity. The sight of lightning crackling through a volcanic eruption is one of the most dramatic displays of nature's fury.

Bioluminescent Bays: Where Water Glows in the Dark In places like Mosquito Bay in Puerto Rico, the water glows with an eerie blue light at night. This bioluminescence is caused by tiny organisms called dinoflagellates that produce light when disturbed. Paddling through the bay at night creates glowing trails in the water, making it one of the most magical and otherworldly experiences one can witness.

Chapter 3: Hidden Geological Wonders

Mysterious Sinkholes: Earth's Sudden Disappearances Sinkholes are one of nature's most abrupt and shocking phenomena. Without warning, the ground opens up and swallows everything above it, from trees to entire buildings. These geological formations occur when underground water erodes limestone, creating cavities that eventually collapse. From the massive sinkholes in Guatemala City to the Dragon's Hole in the South China Sea (the deepest known sinkhole in the world), sinkholes remind us of the hidden forces shaping the Earth beneath our feet.

Lake Natron: A Caustic and Deadly Lake Natron in Tanzania is one of the most inhospitable bodies of water on Earth.

With its high alkaline content, the lake's water can reach temperatures of over 120°F and is so caustic that it can burn the skin and eyes of animals. Birds that land on the lake often become encrusted in salt and die, their bodies perfectly preserved in the lake's toxic waters, creating a macabre display of life and death.

Salar de Uyuni: The World's Largest Natural Mirror The Salar de Uyuni in Bolivia is the world's largest salt flat, stretching over 4,000 square miles. After rain, the flat becomes a gigantic mirror, reflecting the sky in perfect symmetry. This breathtaking phenomenon creates an illusion of walking on air, as the boundaries between sky and ground disappear.

Chapter 4: Unusual Species and Adaptations

The Tardigrade: Earth's Most Resilient Creature Tardigrades, also known as water bears, are microscopic animals that can survive some of the harshest conditions known to life. They've been found in the deep sea, in the frozen tundra, and even in outer space. Tardigrades can endure extreme radiation, boiling heat, freezing temperatures, and the vacuum of space by entering a state of suspended animation. They are living proof of life's ability to persist in the most inhospitable environments.

Glass Frogs: Nature's Transparent Wonders In the rainforests of Central and South America, glass frogs have evolved a unique adaptation: their skin is transparent, allowing you to see their internal organs. This transparency helps them avoid predators, blending seamlessly into their environment while perched on leaves.

Bioluminescent Creatures: Life's Own Light Show In the dark depths of the ocean, bioluminescent organisms light up the underwater world. From the glowing jellyfish to the anglerfish with its bioluminescent lure, these creatures have developed their own natural light sources to hunt, attract mates, or ward off predators. Their eerie glow illuminates the hidden corners of Earth's oceans.

Chapter 5: Human Interaction with Strange Environments

Coober Pedy, Australia: Living Underground In the remote Australian outback, the town of Coober Pedy has adapted to its extreme desert heat by building homes underground. Residents live in "dugouts," which provide natural insulation from the scorching temperatures above. The underground town includes homes, churches, and even hotels, all carved into the rock. This innovative adaptation to a hostile environment shows how humans can live in harmony with nature by using the Earth itself as shelter.

Nomadic Tribes of the Gobi Desert: Thriving in Extremes The Gobi Desert, one of the world's largest and coldest deserts, is home to nomadic tribes that have learned to survive in one of the most extreme environments on Earth. For thousands of years, the Mongolian nomads have adapted to the harsh conditions of scorching summers and freezing winters. They live in portable dwellings known as yurts, which offer protection from the brutal desert winds, and rely on herding animals like camels, goats, and yaks for food, clothing, and shelter. The nomads' deep connection to the land and their resourcefulness have allowed them to maintain a lifestyle that remains virtually unchanged by time.

Life on Tristan da Cunha: The World's Most Remote Inhabited Island Tristan da Cunha is often called the world's most remote inhabited island, located over 1,500 miles from the nearest landmass,

Saint Helena. This isolated volcanic island in the South Atlantic Ocean is home to fewer than 300 people, who live in a small settlement called Edinburgh of the Seven Seas. Cut off from the modern world, the islanders lead a self-sufficient life, farming potatoes, raising livestock, and fishing. Despite the challenges of living in such an isolated environment, the inhabitants of Tristan da Cunha have built a close-knit community that thrives on cooperation and resilience.

Chapter 6: Ecological Mysteries

The Arctic Bermuda Triangle: A Frozen Enigma The Arctic region is home to its own version of the Bermuda Triangle, an area where ships and planes have mysteriously disappeared. Known as the Devil's Sea or the Arctic Bermuda Triangle, this icy expanse has puzzled explorers and scientists for centuries. Some believe the unusual magnetic fields in the region cause navigational equipment to malfunction, while others suggest extreme weather and rogue waves may be responsible for the mysterious vanishings. The Arctic's harsh and unpredictable environment continues to fuel the mystery of this cold and treacherous area.

The Great Pacific Garbage Patch: A Man-Made Island of Waste In the heart of the Pacific Ocean lies the Great Pacific Garbage Patch, a massive accumulation of plastic waste that spans an area twice the size of Texas. This floating island of debris is formed by ocean currents that trap and concentrate plastic and other waste materials. The environmental impact of the Garbage Patch is staggering, as it disrupts marine ecosystems, endangers wildlife, and contributes to the growing crisis of ocean pollution. Despite its grim existence, the patch serves as a reminder of humanity's impact on the planet and the urgent need for environmental action.

The Walking Forests of the Amazon: Trees That Move In the Amazon rainforest, certain trees have developed an unusual adaptation: the ability to "walk." The Walking Palm (Socratea exorrhiza) is a species of tree that can move several centimeters per year to find better sunlight or more stable ground. It achieves this through its stilt-like roots, which grow new roots in the direction of more favorable conditions while the old roots decay. Though slow-moving, this incredible adaptation allows the tree to thrive in a constantly shifting rainforest environment.

Chapter 7: The Weird and Wonderful in Flora

The Corpse Flower: A Giant That Smells Like Death The Titan Arum, more commonly known as the Corpse Flower, is one of the world's largest and rarest flowers, and it's famous for its putrid smell, which resembles rotting flesh. The stench attracts carrion-eating insects, such as flies and beetles, which help pollinate the plant. The Corpse Flower only blooms once every few years, and when it does, it creates a spectacle that draws crowds of curious visitors. This massive flower can grow up to 10 feet tall, and its brief bloom, coupled with its foul odor, makes it one of nature's most bizarre and fascinating plants.

Fungi Superhighways: The Hidden Communication Network of Forests

Beneath the surface of the forest floor lies an underground network of fungi known as mycorrhizal networks, or "the wood-wide web." These fungi form symbiotic relationships with the roots of trees and plants, allowing them to communicate and share nutrients across vast distances. Through this underground network, trees can send nutrients to one another, warn each other of pests, and even support the growth of young saplings. This hidden communication system plays a crucial role in the health and sustainability of forests, and its discovery has revolutionized our understanding of how ecosystems function.

The Dragon's Blood Tree of Socotra: A Tree That Bleeds The Dragon's Blood Tree (Dracaena cinnabari), found on the island of Socotra in Yemen, is one of the world's most unique and otherworldly trees. Its umbrella-shaped canopy provides shade for the arid landscape, and its deep red sap, known as "dragon's blood," has been used for centuries in traditional medicine, dyes, and cosmetics. The tree's unusual appearance and its ability to survive in one of the harshest environments on Earth have made it a symbol of the island's rich biodiversity. Socotra itself is often referred to as the "Galápagos of the Indian Ocean" due to its high number of endemic species, including the Dragon's Blood Tree.

Chapter 8: Unusual Environmental Conservation Efforts

The Global Seed Vault: Humanity's Insurance Policy On the remote island of Svalbard, Norway, deep within a mountain, lies the Svalbard Global Seed Vault, often referred to as the "Doomsday Vault." This facility serves as a backup storage for the world's crop seeds, ensuring that vital plant species are preserved in the event of global disasters, such as climate change, war, or disease. The vault is designed to withstand natural and man-made catastrophes, and its location in the Arctic provides natural refrigeration to keep the seeds viable for centuries. This ambitious conservation project is humanity's insurance policy against a future where food security may be threatened.

Vertical Farming: Growing Food in the Sky

As urban populations continue to grow, space for traditional farming is becoming scarce. Enter vertical farming, a revolutionary agricultural method that involves growing crops in stacked layers, often in urban skyscrapers. Using hydroponic or aeroponic systems, vertical farms can produce food year-round while using less water and land than traditional agriculture. Some cities, like Singapore and New York, are embracing vertical farming as a sustainable solution to feeding their populations in a world where arable land is becoming increasingly limited.

Mushrooms as Nature's Cleanup Crew
Fungi, particularly certain types of mushrooms, are emerging as powerful tools for environmental cleanup. Mycoremediation is a process that uses fungi to break down pollutants, such as oil spills, heavy metals, and pesticides. One species of mushroom, Pleurotus ostreatus, can even break down plastic, offering a potential solution to the growing problem of plastic pollution. Researchers are exploring ways to harness the power of mushrooms to clean up contaminated environments, restore ecosystems, and create sustainable, eco-friendly products.

Conclusion: The Marvel of Earth's Hidden Wonders

As we've journeyed through the obscure and unusual facts about our planet, one thing becomes clear: Earth is a treasure trove of mysteries waiting to be discovered. From the depths of the ocean to the highest mountain peaks, from deserts to rainforests, our biosphere is filled with life forms and phenomena that continue to defy explanation. By understanding these hidden wonders, we gain a deeper appreciation for the complexity and beauty of the natural world, and we are reminded of the importance of protecting the delicate balance that sustains life on Earth.

Dear Reader,

Thank you for joining me on this incredible journey through some of Earth's most extraordinary and hidden wonders. It's a joy to share these fascinating stories, discoveries, and mysteries with those who are as curious about our planet as I am.

Your time and interest in exploring the world's rarest phenomena are greatly appreciated, and I hope this book has sparked a deeper appreciation for the complexity and beauty of the natural world. Together, we can marvel at our planet's diversity and be inspired to protect its precious ecosystems for generations to come.

With gratitude,

Linda Lewis

www.ingramcontent.com/pod-product-compliance
Lightning Source LLC
Chambersburg PA
CBHW040254220526
45473CB00001B/474